FLORA OF TROPICAL EAST AFRICA

SONNERATIACEAE

G. R. Williams Sangai
(East African Herbarium)

Trees or shrubs. Leaves opposite, simple, entire, exstipulate. Flowers ☿, actinomorphic, solitary or 3 together, axillary or terminal. Calyx thick and leathery; tube campanulate; lobes 4–8, valvate. Petals 4–8 or absent. Stamens 12 to numerous, inserted on the calyx, often in several series; filaments free, at length reflexed; anthers reniform or oblong, medifixed, opening lengthwise. Ovary free or adnate to the calyx-tube at the base, 4–many-locular; septa thin; ovules numerous, on thick axile placentas, ascending; style long, simple; stigma capitate. Fruit a berry or a valvate capsule with 4–many locules and numerous seeds. Seeds without endosperm; embryo with short leafy cotyledons.

A small Old World family of two genera, one of which occurs in eastern Africa. Formerly included in the *Lythraceae*.

SONNERATIA

Linn. f., Suppl.: 38 (1781), *nom. conserv.*

Blatti Adans., Fam. Pl. 2: 88 (1763); Niedenzu in E. & P. Pf. III.7: 20 (1892)

Glabrous trees or shrubs with coriaceous leaves. Calyx lobes 6–8, about as long as the tube, valvate in bud. Petals as many as the lobes of the calyx and shorter than them, broad and wrinkled or narrow and smooth. Stamens numerous, in several rows, inserted on a ring at the top of a perigynous sheath. Ovary adnate to the tube of the calyx towards its base, depressed-globose, multilocular; ovules numerous in each locule; style straight; stigma subcapitate. Fruit a multilocular berry, ultimately free from the calyx and stipitate; locules many-seeded. Seeds curved, angular, with a thick crustaceous testa; cotyledons shorter than the terete radicle.

A small genus of five species occurring in mangrove-swamps from eastern Africa to the western Pacific.

S. alba *Sm.* in Rees Cycl. 33, No. 2 (1816); DC., Prodr. 3: 231 (1828); C.B. Cl. in Fl. Brit. India 2: 580 (1879); Engl., V.E. 3(2): 655–656 (1921); T.T.C.L.: 592 (1949); Backer & van Steenis, Fl. Males., ser. 1, 4(3): 285, fig. 3/b (1951); E.P.A. 612 (1959); K.T.S.: 539, fig. 98 (1961). Type: Indonesia, Amboina, *Rumphius*, illustration of *Mangium caseolare album* in Herb. Amb. 3: 111, t. 73 (1743)

An evergreen shrub or usually a small tree, 3–15(–20) m. tall, with many stout finger-like pneumatophores and spreading branches; bark dark grey or greyish-brown, rough and fissured. Leaf-blades yellow-green, obovate, oval, or almost round, 3·8–12·5 cm. long, 1·7–9 cm. wide, rounded or emarginate at the apex, cuneate at the base; lateral nerves 11–14 on either side of the prominent midrib; petiole stout, 3–15 mm. long. Flowers scented, solitary or 3 together at apices of the shoots. Buds oblong, narrowed at base

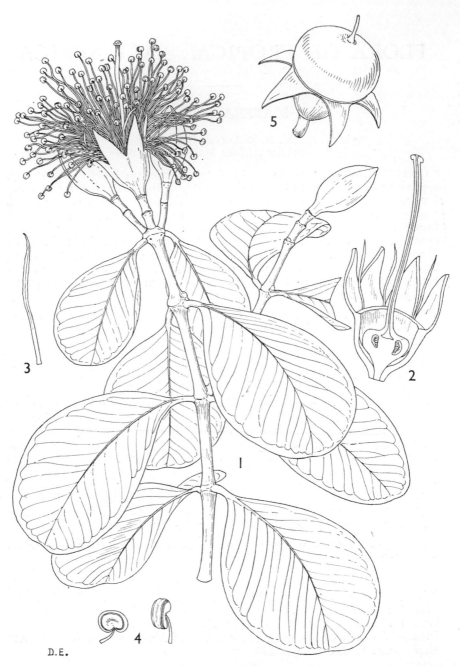

D.E.

FIG. 1. *SONNERATIA ALBA*—**1**, flowering branch, × ⅔; **2**, longitudinal section of flower with stamens removed, × 1; **3**, petal, × 2; **4**, anthers, two aspects, × 3; **5**, fruit, × ⅔. 1, 3, 4, from *Drummond & Hemsley* 3234; 2, from a drawing by Kirk; 5, from *Mogg* 28730.

and apex.   Calyx green, 2·6–3·5 cm. long; tube campanulate, angular, the angles alternating with the calyx-lobes; lobes 6–8, magenta-pink inside, green outside, shorter or longer than the tube, 1·2–2 cm. long and 4·5–9(–12) mm. wide, at first more or less erect, later reflexed in fruit.   Petals white or tinged with magenta-pink, soon deciduous, strap-shaped, inconspicuous and closely resembling the filaments, 13–20 mm. long and 0·5–1·25 mm. wide. Stamens numerous, the filaments showy, white, inflexed in bud.   Ovary 14–18-locular; style green, 4·5–5·9 cm. long; stigma capitate, 2·5–3 mm. wide.   Fruit obconic-turbinate, 2·3–3 cm. long and 3·1–4 cm. wide, green, crowned by the style base.   Ripe seeds not seen.   Fig. 1.

KENYA.   Kwale District: Kikambala, June 1954, *Irwin* 24 in *C.M.* 22259 !; Mombasa I., Tudor House Beach, 7 Aug. 1965, *G. R. Williams Sangai* 844 !; Kilifi District: Kilifi Creek, July 1939, *C. G. van Someren* Sh 91 !

TANGANYIKA.   Tanga District: 11 km. NE. of Pangani, Kigombe Beach, 11 July 1953, *Drummond & Hemsley* 3234 !; Rufiji District: Rufiji delta near Salale, 25 Feb. 1900, *Busse* 402 ! & Mafia I., Kanga, 15 Aug. 1937, *Greenway* 5113 !; Kilwa District: Kilwa Kisiwani, 2 June 1906, *Braun* 1180 !

ZANZIBAR.   Zanzibar I., Unguja Ukoo, 5 Feb. 1929, *Greenway* 1355 !; Pemba I., Chake Chake, 17 July 1929, *T. C. Vaughan* 404 !

DISTR.   **K**7; **T**3, 6, 8; **Z**; **P**; Mozambique, Somali Republic (S.), N. Madagascar, Comoro Is., Seychelles, SE. Continental Asia and Andamans to N. Australia, S. Riu Kiu Is., Micronesia, Solomon Is., New Hebrides and New Caledonia

HAB.   A common outer fringe constituent of mangrove-swamps in very deep mud; sea-level

SYN.   *S. mossambicensis* Klotzsch in Peters, Reise Mossamb. Bot. 1: 66, t. 12 (1861). Type: Mozambique, 14–18°S., *Peters* (B, holo. †, K, iso.)
[*S. acida* sensu Hiern in F.T.A. 2: 483 (1871), *non* Linn.f.]
[*S. caseolaris* sensu P.O.A. C: 286 (1895); U.O.P.Z.: 449, fig. et auct. al. afr. *non* (L.) Engl.]
S. acida Linn. f. var. *mossambicensis* (Klotzsch) Mattei in Boll. Ort. Palermo 7: 108 (1908)

NOTE.   The type of *S. mossambicensis* Klotzsch can scarcely be anything but *S. alba* Sm., but the fruit is drawn with erect calyx-lobes and petals are stated to be absent. I have followed Backer and van Steenis in considering it a synonym.

INDEX TO SONNERATIACEAE